Geometry and Fractions with Pattern Blocks

Problem-Solving Activities
Grades 3-6

by
Barbara Bando Irvin, Ph.D.

Table of Contents

Introduction

Geometry and Fractions with Pattern Blocks contains activities to help students in grades 3-6 reinforce the basic mathematical concepts of geometry and fractions while applying problem-solving strategies. Many of the activities may be integrated in your day to day mathematics curriculum.

The activities in this book are designed to be used with a set of Pattern Blocks (LER 134). Pattern blocks consist of six shapes and six colors: yellow hexagons, red trapezoids, blue and tan rhombuses, green triangles, and orange squares. A set of Overhead Pattern Blocks (LER 640) may be used to demonstrate activities and problems on the overhead projector or used to show solutions.

A pattern block blackline master is provided on page 7 for students who wish to make their own set. Note that for every hexagon there are two trapezoids, three blue rhombuses, and six triangles to enable students to make the hexagon with other shapes. Color and cut out several copies of this page so that students can make very large geometric shapes as well as whole pages of patterns. Hexagon grid paper, isometric grid paper (solid and dotted lines), and one-inch grid paper are provided on pages 8-11 to help students discover geometric relationships, invent fraction problems, and create "quilt" patterns. Selected solutions can be found at the back of the book.

Overview of Content

The activities in this book were developed using NCTM's *Curriculum and Evaluation Standards for School Mathematics* (1989) as a guide.

Special attention was paid to K-4 Standard 9: *Geometry and Spatial Sense* and Standard 12: *Fractions and Decimals*; and 5-8 Standard 6: *Number Systems and Number Theory* and Standard 12: *Geometry*.

The mathematical content is presented in two parts with the following goals:

Geometry

• To describe, model, draw, and classify shapes.
• To explore congruence and similarity of shapes.
• To predict results by combining, subdividing, or changing shapes.
• To measure, describe, and classify angles.
• To find reflections in a mirror.
• To recognize symmetrical shapes and locate lines of symmetry.
• To learn about transformational moves.
• To discover perimeter and area patterns.

Fractions

• To name fractional parts of a region or set.
• To name and represent unit, proper, and equivalent fractions.
• To discover ways to make sums of $\frac{1}{2}$ and 1.
• To add fractions with like and unlike denominators.

Developing Problem-Solving Skills

Problem-solving skills such as using a model, drawing a picture or diagram, making an organized list, guess-and-check, working backwards, looking for patterns, and using logical reasoning may be applied to many activities. These skills can then be used to find multiple solutions to every day problems.

Encourage students to make geometric shapes with pattern blocks. Students may make rectangles, pentagons, and hexagons. Pattern blocks can be used in the same manner as tangrams to create designs and pictures. The examples below show a fat cat, bird, camel, candle, house, flower in a vase, wrench, sunburst, tree, letter B, and numeral 2.

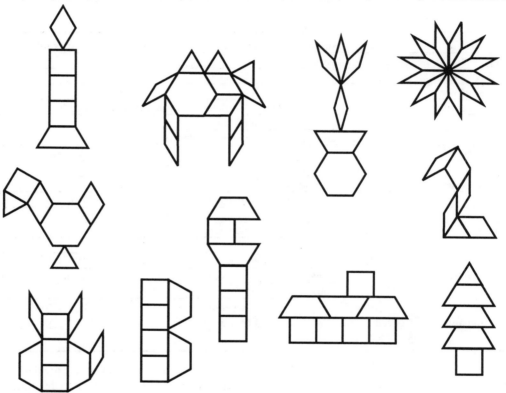

Using *Geometry and Fractions with* Pattern Blocks

Each section begins with *Teaching Notes* to provide an overview of the content and suggestions for classroom use. *Teaching Notes* includes:

Getting Ready: An activity to introduce the mathematical concepts in the section. This may consist of a manipulative activity or an oral exercise.

Activity Teaching Notes: The objective and a description of each blackline master activity.

Pattern Block Activities: The blackline masters for each activity described in the *Teaching Notes*.

Students should be encouraged to work cooperatively in pairs or small groups to find multiple solutions for each activity. Urge them to create their own problems and puzzles to challenge one another. Pattern blocks are a hands-on, motivational tool that work with an array of mathematical concepts, relationships, and creations.

Getting Started

Allow students to familiarize themselves with a set of pattern blocks. Discuss the number of sides, color, and shape of each block. Observe and listen to students as they manipulate the blocks. Some students will discover that three green triangles will cover a red trapezoid or that six green triangles cover a hexagon. Ask students to share their discoveries with their classmates.

About Pattern Blocks

This chart contains a number of facts that are helpful while presenting lessons or answering students' questions. It also includes important vocabulary terms that should be defined and discussed throughout the activities.

Shape	triangle	square	rhombus	rhombus	trapezoid	hexagon
Color	green	orange	blue	tan	red	yellow
Number of sides	3	4	4	4	4	6
Number of vertices	3	4	4	4	4	6
Polygon category	triangle	quadrilateral	quadrilateral	quadrilateral	quadrilateral	hexagon
Equivalent sides	all 3 sides	all 4 sides	all 4 sides	all 4 sides	3 of 4 sides	all six sides
Congruent angles	all 3 angles	all 4 angles	opposite angles	opposite angles		all 6 angles
Polygon type	equilateral*	regular*			isosceles	regular
Parallel sides		2 pairs	2 pairs	2 pairs	1 pair	3 pairs
Opposite sides		equivalent	equivalent	equivalent		equivalent
Length of sides	1 in. each	1 in. each	1 in. each	1 in. each	1 in.,2 in.	1 in. each
Perimeter	3 in.	4 in.	4 in.	4 in.	5 in.	6 in.
Area (approx.)	0.43 sq. in.	1 sq. in.	0.86 sq. in.	0.43 sq. in.	1.3 sq. in.	2.6 sq. in.
Symmetry	3 lines	4 lines	2 lines	2 lines	1 line**	6 lines
Interior angles	60° each	90° each	60° and 120°	30° and 150°	60° and 120°	120° each
Angle names	acute	right	acute, obtuse	acute, obtuse	acute, obtuse	obtuse
Sum of angles	180°	360°	360°	360°	360°	720°
Fractional relationships	$\frac{1}{6}$ of yellow $\frac{1}{3}$ of red $\frac{1}{2}$ of blue		$\frac{1}{3}$ of yellow $\frac{2}{3}$ of red		$\frac{1}{2}$ of yellow	

* The term "regular polygon" refers to a polygon that is equilateral and equiangular. A "regular" triangle is usually called an equilateral triangle. ** An isosceles trapezoid has one line of symmetry but this cannot be generalized to any trapezoid.

Although the blue and tan blocks may resemble a "diamond," insist that students call them by their correct name, "rhombus." The blue and tan rhombuses are special parallelograms because they have congruent sides. The square is a special parallelogram because its sides and its angles are congruent. It is also a special rectangle because all four sides have the same length.

All pattern blocks sides have a length of one inch except for one side of the trapezoid. Its length is twice as long or two inches. Because of the relationship between the lengths of the sides of each block, geometric shapes and patterns may be formed using all of the pattern blocks. The green triangles, blue rhombuses, red trapezoids, and yellow hexagons work especially well together since they contain only 60 degree and 120 degree angles.

Set of Pattern Blocks

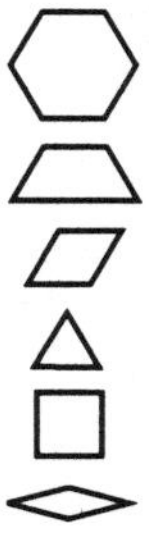

◇ Yellow
△ Red
▱ Blue
△ Green
▢ Orange
◇ Tan

Hexagon Grid Paper

Geometry and Fractions with Pattern Blocks
©Learning Resources, Inc.

Isometric Grid Paper

Dotted Isometric Grid Paper

One Inch Grid Paper

Geometry with Pattern Blocks

Vocabulary

triangle, square, rectangle, pentagon, hexagon, octagon, polygon, quadrilateral, rhombus, parallelogram, area, congruent, perimeter, symmetry

Getting Ready

Review the names of various geometric shapes (triangle, square, rectangle, pentagon, hexagon, octagon, and so on) with students. Ask them to make one of the shapes and tell which pattern blocks were used. After students make several shapes, have them trace and color their shapes on the dotted grid paper (page 10) and cut them out. Challenge them to sort and classify the shapes in terms of quadrilaterals, regular polygons, and other attributes.

Pattern Block Activities

Sort and Classify Pattern Blocks *(pages 15-16)*

Several geometric shape names are given on these pages to help students sort and classify the pattern blocks with regard to sides and angles. *Quadrilaterals* include the *square,* both *rhombuses,* and the *trapezoid* since all of these blocks have four sides. The *hexagon, triangle,* and *square* are *regular polygons* because they are *equiangular* and *equilateral.* Both rhombuses and the square can be classified as *parallelograms* since they are closed figures made from the intersection of two pairs of parallel lines. They are rhombuses since they also have congruent sides. Only the square can be classified as a *rectangle* because it contains right angles. Sorting activities help students to find definitions in a broad sense and as well as in specific terms. For example, all squares are parallelograms but not all parallelograms are squares.

Make Other Shapes *(pages 17-18)*

Ask students to give definitions for the shapes used on these two pages before doing them. Include *rectangles, pentagons, hexagons, parallelograms,* and *octagons.* Insist that students specify the number of sides, the size of the angles, and other attributes for each type of polygon. The number of blocks and colors are specified for each shape that is to be formed. Urge students to make several more of these shapes, draw and color them, discuss them, and then display them on the bulletin board.

Congruent Pattern Block Shapes *(pages 19-23)*

Congruent figures have the same size and shape. Students are to re-create some of the actual pattern block shapes using smaller pattern blocks. On page 20, a shaded triangle and a rhombus are shown on dotted isometric grid paper to help students find shapes congruent to those shaded. Have students cover these shaded figures with pattern blocks and then find congruent figures using different colored blocks. Students can organize their work by making a chart of solutions as suggested at the bottom of the page.

Outlines of a trapezoid and a hexagon are given with a chart to keep track of solutions. Encourage students to fill in the trapezoid outline on page 22 with a set of blocks consisting of three colors. Ask them to use the greatest and fewest number of pattern blocks possible. Since there are numerous solutions to each of these pages, students are only required to find those involving a specified number of colors. However, at the bottom of each page, students are challenged to find additional solutions.

Larger Pattern Blocks *(pages 24-25)*

Show how each pattern block can be made proportionally larger. On each page a pattern block is shown in the left column and a larger similar pattern block is shown to its right. Similar figures have the same shape and are proportional in size. Students are asked to make an even larger version of each block on the back of their worksheet. Encourage students to record their solutions. Have them place, trace, and color the blocks for each similar figure.

Similar Pattern Block Shapes *(pages 26-27)*

On page 26, two triangular figures are shaded on the grid paper to show students a pair of similar triangles. Challenge students to find and record three more similar triangles on the grid paper. A pair of similar parallelograms are shown on page 27 without the aid of a grid. Provide grid paper to those students who may need it in order to find other similar parallelograms. Ask students about the relationship between the long and short sides of the parallelograms.

Area and Perimeter of Similar Figures *(pages 28-29)*

Several mathematical concepts are integrated into the activity on page 28 -- *similar* triangles, *area*, *perimeter*, and number patterns. Remind students that the *area* of a figure is the number of unit blocks covering the figure. *Perimeter* is the number of units around the figure. Square orange blocks are used as units on page 28, green triangles are used as unit blocks on page 29. Challenge students to discover the relationship between the area and perimeter of: 1) rectangles using the orange square as the unit; 2) isosceles trapezoids using the red block as the unit; 3) and regular hexagons using the yellow block as the unit. Provide square and isometric grid paper for students to complete these challenges.

Symmetry of Pattern Block Shapes *(pages 30-31)*

Discuss the concept of *symmetry* with students. Encourage students to work with paper cutouts to find lines of symmetry. On page 30, students are to predict how many lines of symmetry there are for each block and then color, cut out, and fold the pattern block cutouts to find the actual number of lines of symmetry. On page 31, students are to predict and then find the lines of symmetry for various pattern block shapes.

Reflections in a Mirror *(page 32-33)*

Help students become familiar with finding reflection images in a mirror and recognizing images that have the same size and shape (congruent). This activity helps students visualize mirror images and reinforces lines of symmetry for a given figure. Students will need a small one-sided mirror. Provide some guided practice in placing the mirror on the design to see each of the figures shown on page 32. Record the placement of the mirror (line of symmetry) and the direction the mirror is pointing. On page 33, students are shown a pair of hexagons. The mirror line is provided on the hexagon design on the left. Students must position the mirror on that line and then draw and color the resulting design on the hexagon outline at the right.

Transformational Moves *(pages 34-37)*

The activities *Glide Slides, Flip Trips,* and *Learn Turns* each deal with transformational moves including translations, reflections, and rotations. These three activity pages along with page 37 will help students gain an appreciation for the artistic aspect of applying geometric concepts. Encourage students to discuss their motif (basic design that has been moved by a slide, flip, or turn) and patterns created.

Angle Measure *(pages 38-40)*

On page 38, students are to find the measure of certain angles for each pattern block. Remind them that a complete rotation is 360 degrees. Using deductive reasoning and division, students should be able to determine the measure of each indicated angle without the aid of a protractor. Note that the pattern blocks in each figure are rotated around a point in the middle of the figure. On pages 39 and 40, students are to find the sum of the angles for each pattern block and for other polygons formed by pattern blocks.

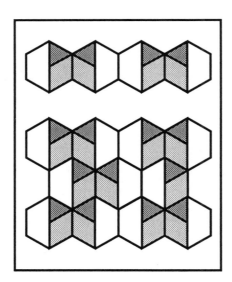

Describe Pattern Blocks

Write a short description for each vocabulary term. Then trace and color the appropriate pattern blocks for each term. Some of the blocks will be used more than once.

Triangle   **Hexagon**

Quadrilateral

Regular Polygons

Pattern Block Quadrilaterals

Write a short description for each vocabulary term. Then trace and color the appropriate pattern blocks. Some of the blocks will be used more than once.

Parallelogram   **Rhombus**

Rectangle   **Square**

Which pattern blocks could be described by all four terms above?

_____

_____

Which pattern blocks could be described by only two terms above?

_____

_____

Geometry and Fractions with Pattern Blocks
©Learning Resources, Inc.

Make Other Shapes

Make three different rectangles. Trace and color your solutions.

Use two blocks.   Use six blocks.

Use three blocks.

Make three different pentagons with the specified colors. Trace and color your solutions.

Use two blocks.   Use three blocks.   Use four blocks.
(Green, Orange)   (Green, Tan)   (Green, Tan, Orange)

Make three different hexagons with the specified colors. Trace and color your solutions.

Use three blocks.   Use three blocks.   Use three blocks.
(Tan, Blue)   (Green, Orange)   (Yellow, Blue)

Make Parallelograms and Octagons

Make three different parallelograms with the specified colors. Trace and color your solutions.

Use two blocks.
(Red, Green)

Use six blocks.
(Red, Green)

Use five blocks.
(Yellow, Red, Green)

Make two different octagons with the specified colors. Trace and color your solutions.

Use four blocks.
(Yellow, Tan, Orange)

Use seven blocks.
(Red, Blue, Orange)

Can you make other parallelograms and octagons?

Geometry and Fractions with Pattern Blocks
©Learning Resources, Inc.

Congruent Pattern Block Shapes

Make the blue shape another way.

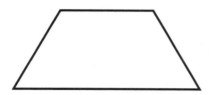

Make the red shape two other ways.

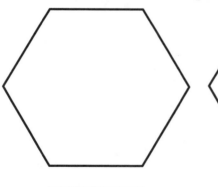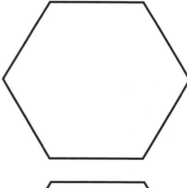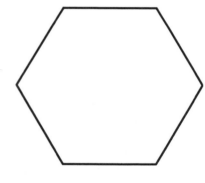

Make the yellow shape five others ways.

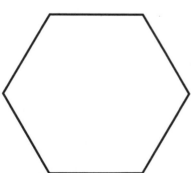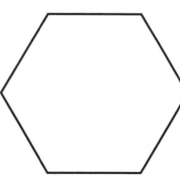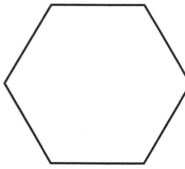

Are there other ways to make the yellow shape using three or four blocks?

Congruent Triangles

Use the dotted grid paper to make triangles congruent to the one shaded below. Trace and color your solutions on the grid below.

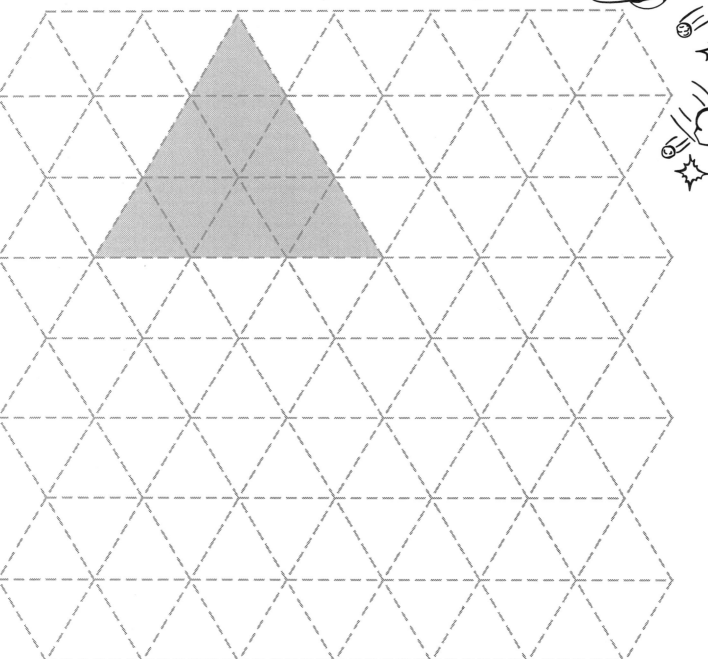

Make a chart of the ways you recreated the shaded triangle. A sample chart is shown at the right.

Find at least five solutions.

Number of blocks	Number of each color			
	Yellow	Red	Blue	Green

Geometry and Fractions with Pattern Blocks
©Learning Resources, Inc.

Congruent Rhombuses

Use the dotted grid paper to make rhombuses congruent to the one shaded below. Trace and color your solutions on the grid below.

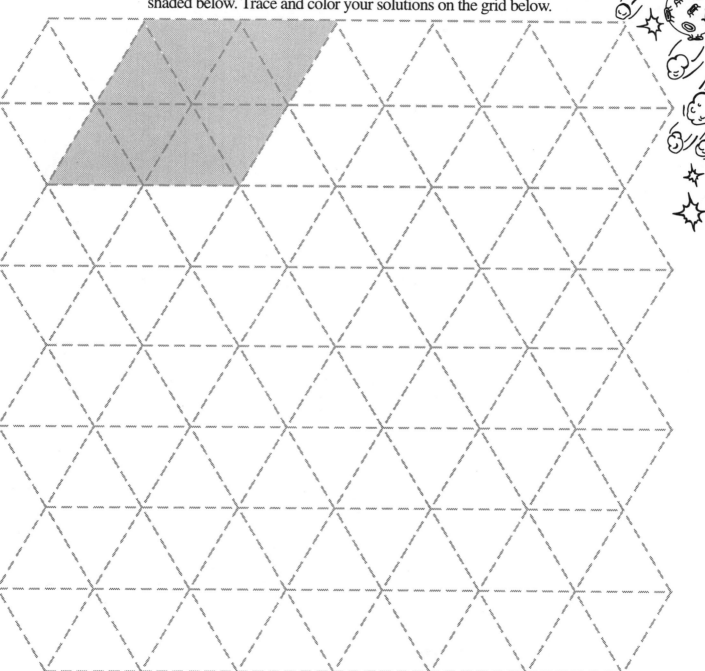

Make a chart of the ways you recreated the shaded rhombus. A sample chart is shown at the right.

Find at least eight solutions.

Number of blocks	Number of each color			
	Yellow	Red	Blue	Green

Congruent Trapezoids

Find eight trapezoids congruent to the figure below. Use three or four colors. Record your solutions in the chart.

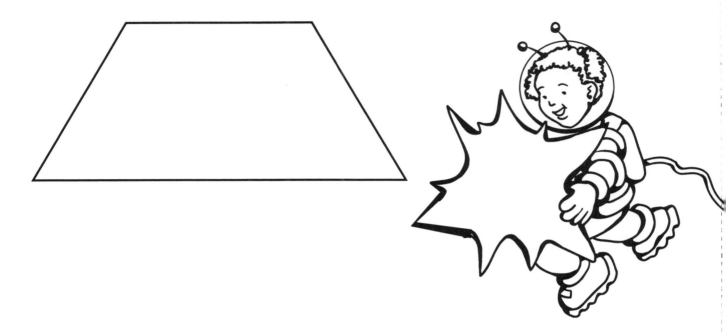

Number of blocks	Number of each color			
	Yellow	Red	Blue	Green

List ways to make the trapezoid above using one or two colors.

Geometry and Fractions with Pattern Blocks
©Learning Resources, Inc.

Congruent Hexagons

Find eight hexagons congruent to the figure below. Use only three colors. Record your solutions in the chart.

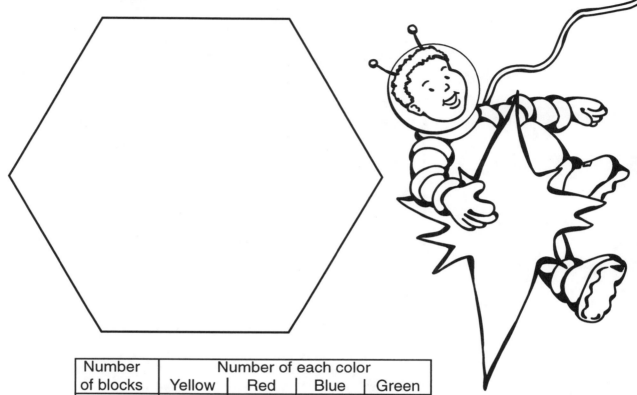

| Number | Number of each color | | | |
of blocks	Yellow	Red	Blue	Green

List ways to make the hexagon above using one, two, or four colors.

Larger Pattern Blocks

Cover each shape with pattern blocks. Make a much larger shape on the back of this paper. Trace and color the blocks on the larger shapes to show your solutions.

Pattern Block   **Larger Shape**

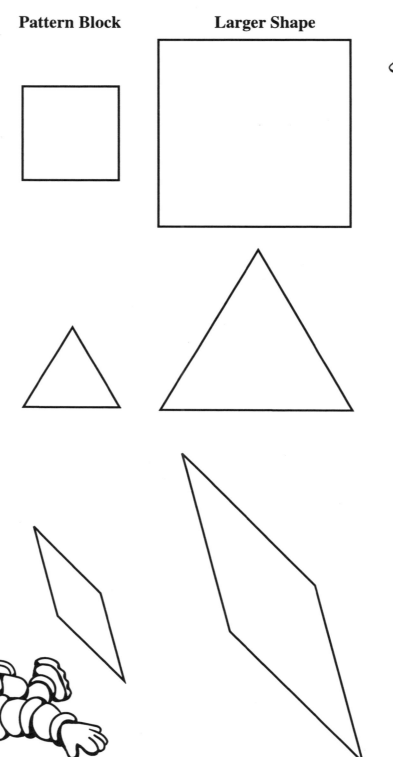

More Larger Pattern Blocks

Cover each shape with pattern blocks. Make a much larger shape on the back of this paper. Trace and color the blocks on the larger shapes to show your solutions.

Pattern Block　　　　　**Larger Shape**

Similar Triangles

Use the dotted grid paper to make three triangles similar to the two that are shaded below. Trace and color the triangles on the grid below to show your solutions.

Similar Parallelograms

Look at the two similar parallelograms below. On the back of this paper make two larger parallelograms that are also similar. Trace and color your solutions.

Find two more parallelograms similar to those above. What do notice about the relationship between the long and short sides of the parallelograms above?

Perimeter and Area of Similar Squares

Cover the shaded square on the grid with a pattern block. Next, cover the grid around the shaded square to make a two by two-unit square, a three by three square, etc. Record the results in the chart.

● What number pattern do you notice for the area and the perimeter?

Shape	Area	Perimeter
1 by 1	1	4
2 by 2	4	___
3 by 3	___	___
4 by 4	___	___
5 by 5	___	___
6 by 6	___	___

Geometry and Fractions with Pattern Blocks
©Learning Resources, Inc.

Perimeter and Area of Similar Triangles

Cover the shaded triangle on the grid with a pattern block. Next, cover the grid around the shaded triangle to make triangles with two, three, four units etc. on each side. Record the results in the chart as you complete each new triangle.

● What number pattern do you notice for the area and the perimeter?

Each Side	Area	Perimeter
1 unit	1	3
2 units	4	6
3 units	___	___
4 units	___	___
5 units	___	___
6 units	___	___

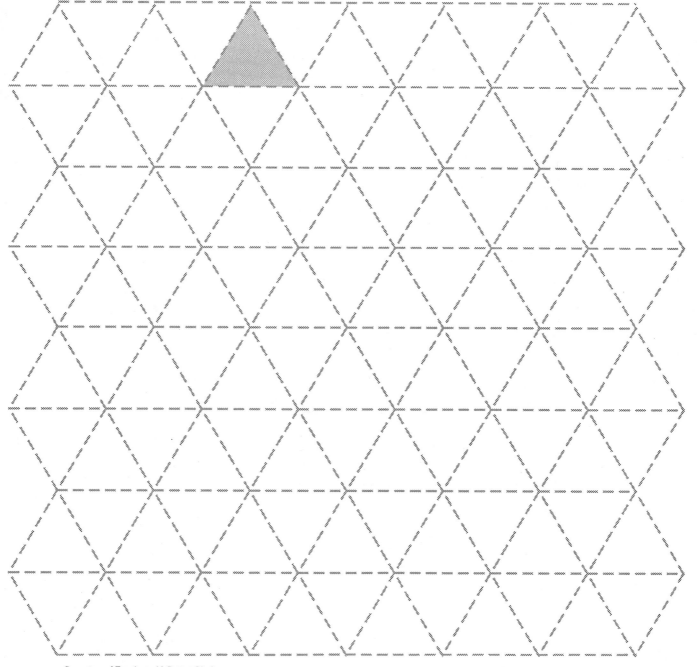

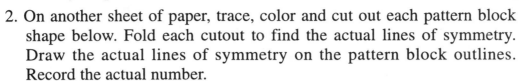

Symmetry of Each Pattern Block

1. Look at each shape. Predict the number of lines of symmetry. Write your prediction below.

2. On another sheet of paper, trace, color and cut out each pattern block shape below. Fold each cutout to find the actual lines of symmetry. Draw the actual lines of symmetry on the pattern block outlines. Record the actual number.

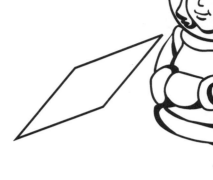

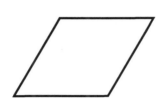

Prediction _____   Prediction _____   Prediction _____

Actual _____   Actual _____   Actual _____

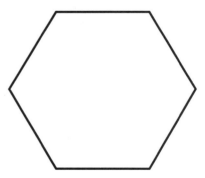

Prediction _____   Prediction _____   Prediction _____

Actual _____   Actual _____   Actual _____

Name...**Date**..............

Symmetry of Shapes

1. Look at each shape. Predict the number of lines of symmetry. Write your prediction below.

2. On another sheet of paper trace, color and cut out each shape. Fold each cutout to find the actual lines of symmetry. Draw the actual lines of symmetry on the outlines. Record the actual number.

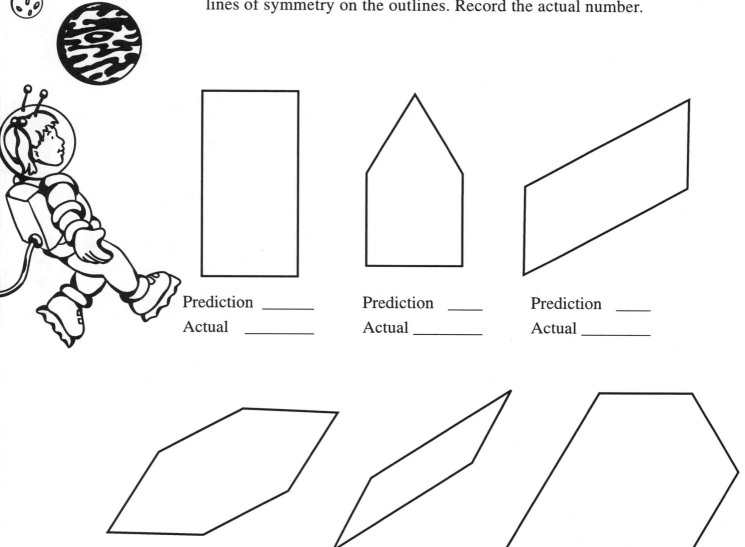

Prediction _____   Prediction _____   Prediction _____
Actual _____   Actual _____   Actual _____

Prediction _____   Prediction _____   Prediction _____
Actual _____   Actual _____   Actual _____

Reflections in a Mirror

Materials Needed: A red, blue, and green pattern block and one-sided mirror.

Place the three blocks as shown on the hexagon at the left. Then move the mirror over or around the hexagon to find the design shown at its right. Draw where the mirror should be placed and which direction it should face. Color the designs to show the results. **Example:**

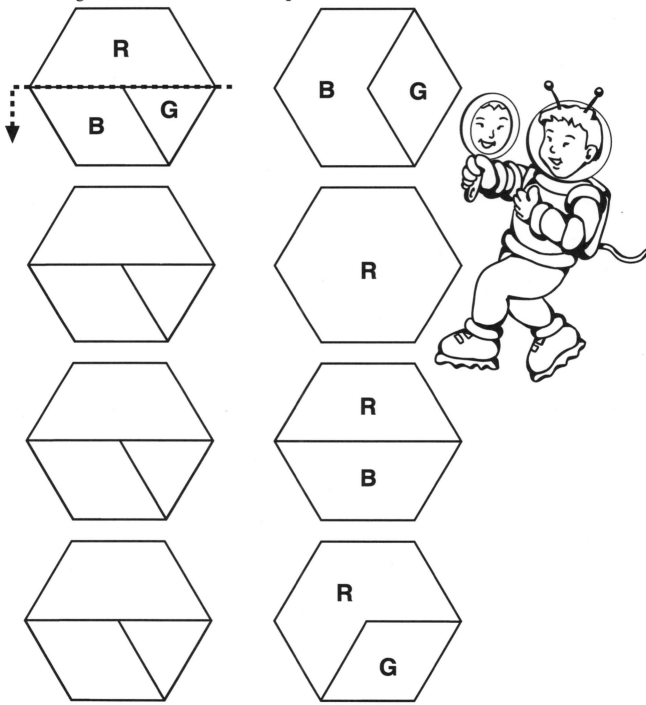

Find more reflection images using the same red-blue-green design above.

Geometry and Fractions with Pattern Blocks
©Learning Resources, Inc.

More Reflections in a Mirror

Materials Needed: A red, blue, and green pattern block and one-sided mirror.

Place the three pattern blocks as shown on the hexagon at the left. Then place a mirror on the line indicated. Draw the resulting design and color it on the hexagon outline.

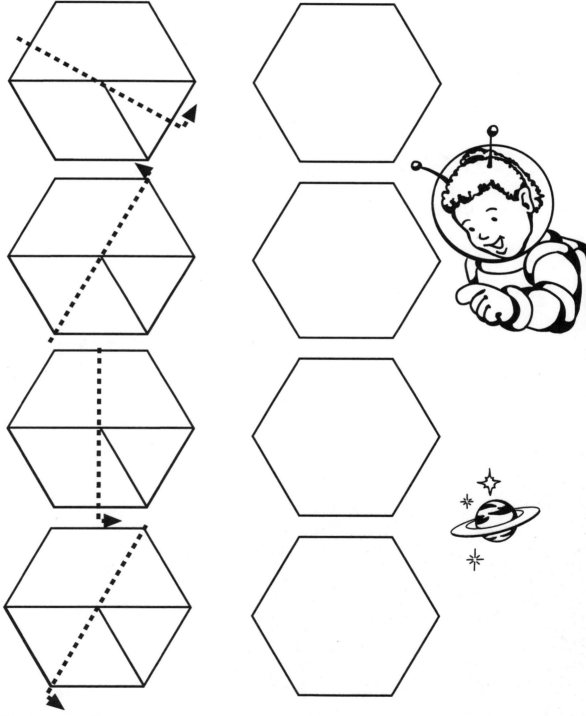

Find more reflection images using the same red-blue-green design above.

Glide Slides

Create a basic two-color or three-color design in the first hexagon. Slide the design to the right onto the next three hexagons. Color the shapes to show the design.

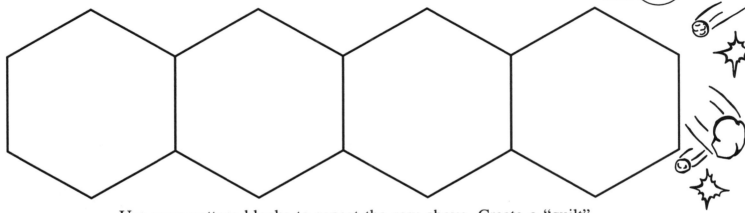

Use your pattern blocks to repeat the row above. Create a "quilt" pattern on the grid below. Color the grid to show the design.

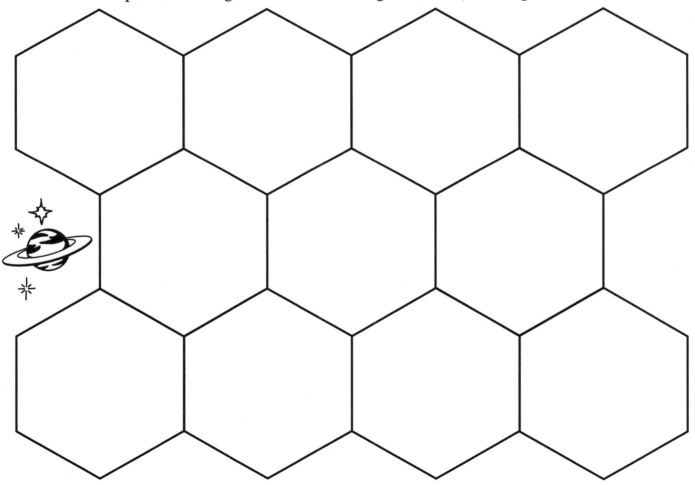

Geometry and Fractions with Pattern Blocks
©Learning Resources, Inc.

Flip Trips

Use pattern blocks to create example shown. Notice the design is flipped on its right side to continue the pattern.

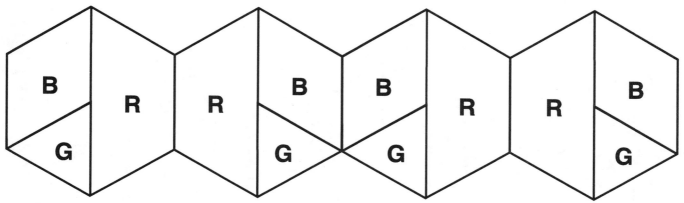

Create two additional flip patterns using 2 or 3 color designs. Color the pattern.

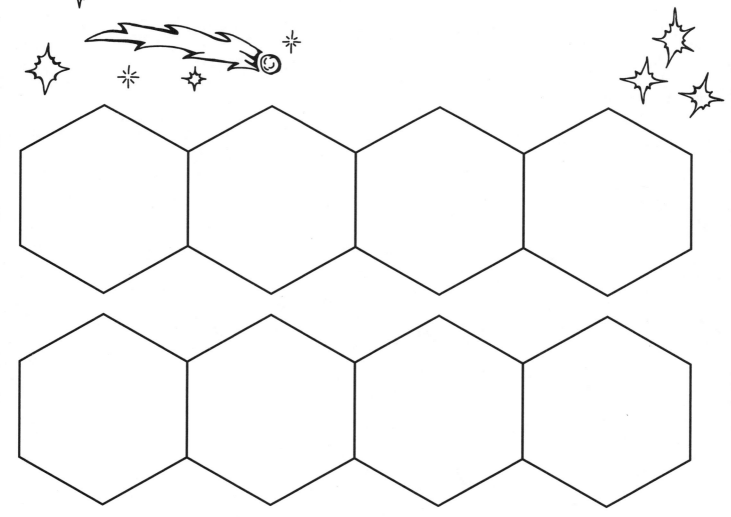

Learn Turns

Use pattern blocks to create example shown. Notice the design is rotated 60 degrees each turn. Color the design.

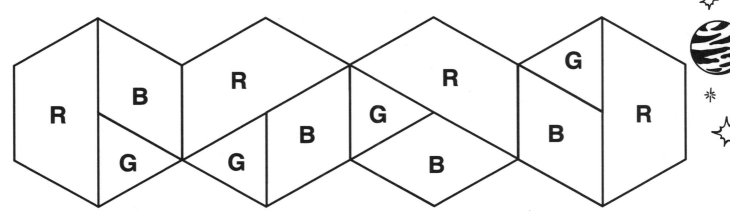

Rotate the design in the first hexagon 120 degrees each turn. Color the design.

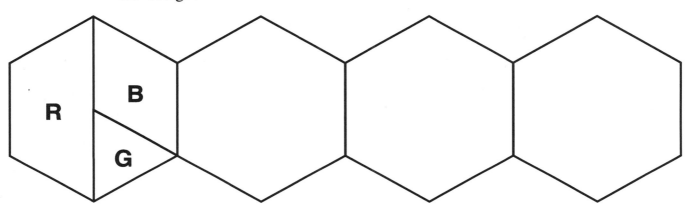

Rotate the design in the first hexagon 180 degrees each turn. Color the design.

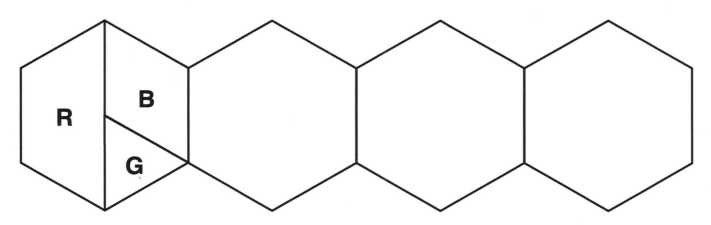

Pattern Block Designs

Use a two-color or three-color design and a slide, flip, or turn to make a pattern block design on this grid paper.

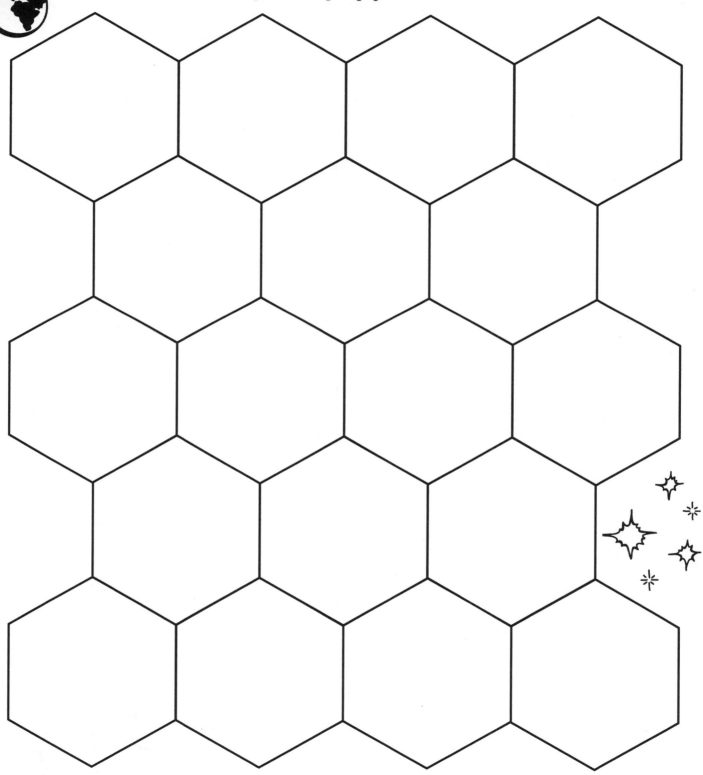

Angle Measure

Use pattern blocks to make each shape. In each shape, the pattern blocks are rotated around a dot in the middle of the figure. A complete rotation is 360 degrees. Find the angle measure for each rotation without using a protractor.

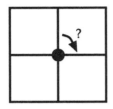

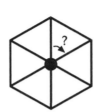

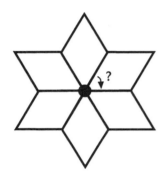

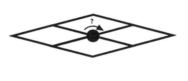

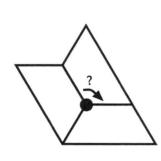

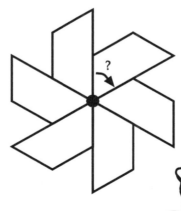

Geometry and Fractions with Pattern Blocks
©Learning Resources, Inc.

Name..Date...............

Pattern Block Sums of Angles

Write in the number of degrees for the vertex angles of each pattern block. Then find the sum of the interior angles for each pattern block.

 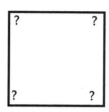 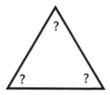 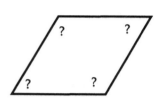

Sum _____   Sum_____   Sum_____

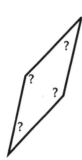 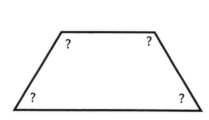 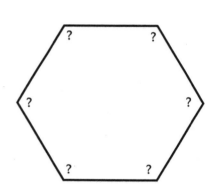

Sum _____   Sum_____   Sum_____

1. What is the sum of the angles of a triangle? _____

2. What is the sum of the angles of a quadrilateral? _____

3. What is the sum of the angles of a hexagon? _____

4. What do you think is the sum of the angles of a pentagon? _____

5. What do you think is the sum of the angles of an octagon? _____

Geometry and Fractions with Pattern Blocks
©Learning Resources, Inc.

Angles Sums of Polygons

Use pattern blocks to make each of the shapes below. Write in the number of degrees for each vertex angle. Find the sum of the angles for each shape.

?   ?

?   ?

Sum _____

?   ?

?   ?

?   ?

Sum _____

?   ?

?   ?

?   ?

?   ?

Sum _____

?   ?

?   ?

?   ?

?   ?

?   ?

Sum _____

?   ?

?   ?

?   ?

?   ?

Sum _____

?

?   ?

?   ?

?

Sum _____

Geometry and Fractions with Pattern Blocks
©Learning Resources, Inc.

Fractions with Pattern Blocks

Vocabulary

unit fraction, proper fraction, numerator, denominator, concentric

Getting Ready

Discuss fractional parts of common things such as a candy bar, pizza, an apple, or a batch of cookies. Emphasize that a fractional part of a region or set is a "fair share" of it. For example, a pizza must be cut into six equal-sized pieces in order for six people to have a fair share. Discuss the difference between a *unit fraction* and a *proper fraction*. In the pizza that is cut into six equal-sized pieces, each piece represents a unit fraction of the whole pizza. A proper fraction is a fraction whose numerator is less than its denominator.

Consider fractional parts of a set using your class as an example. Use students' attributes such as blue eyes, brown hair, and so on, to find fractional parts of the class. For example, if eight students in a class of twenty-four wear glasses, then $\frac{8}{24}$ or $\frac{1}{3}$ of them wear glasses. Use estimation when appropriate. For example, if eleven students have brown hair, the $\frac{11}{24}$ or about $\frac{1}{2}$ of them have brown hair.

Pattern Block Activities

Parts of a Set *(pages 44-47)*

The first four pages of this section provide problem-solving activities involving fractional parts of a set. These activities give students the opportunity to deal with items of different colors and shapes rather than one shape in many colors or one color in many shapes or sizes. This reinforces the use of geometric terms and concepts. On pages 44 and 45, a set of pattern blocks is given and students must find the fractional parts of the set. The activities on pages 46 and 47 are more challenging in that students are given an incomplete set of pattern blocks and must determine which ones are missing.

Fractional Parts of Each Block *(page 48)*

The unit value of the red, blue, and green blocks change with respect to the region that it is in. Encourage students to discuss fractional relationships.

Parts of a Hexagon *(pages 49-50)*

On page 49, students are to cover the shaded portions of each hexagon with green, blue, or red blocks to show fractional parts of the hexagon. Page 50 is more challenging in that students can use two, three, four, five, or six blocks and up to three colors to fill the outlines of each hexagon shown. An example is shown for the first hexagon figure. Since the figures must be completely covered, each of the color portions of the hexagon becomes a fractional part that adds up to the sum of one.

Fractional Parts of a Region *(page 51)*

The problems on this page provide students with the opportunity to work with the orange and tan blocks in addition to the red, blue, green, and yellow blocks. Students are to show the fractional parts indicated using the pattern blocks and then trace and color the blocks to show the solutions. Urge them to make up several more fraction problems similar to these using all six colors of the pattern blocks.

Sums of One *(pages 52-53)*

Sums of One on a Triangle and *Colorful Fraction Hexagons*, utilize the same skills as page 50 in order to find the fractional parts of each figure. Students may wish to make a chart to organize their work in order to find all the solutions for both pages. The problems on these two pages are more difficult because the outline is not that of the yellow block and will not have denominators dealing with halves, thirds, and sixths.

On page 52, the fractions involve halves and fourths for the three triangles at the top of the page and thirds, sixths, and ninths for the six larger triangles. For example, if one of the larger triangles is covered with one green block, one blue block, and two red blocks, then the addition sentence for it could be $\frac{1}{9} + \frac{2}{9} + \frac{2}{3} = 1$ because it shows $\frac{1}{9}$ green, $\frac{2}{9}$ blue, and $\frac{2}{3}$ (or $\frac{6}{9}$) red.

After working with two or three of the hexagon outlines on page 53, students will soon discover that the green block is $\frac{1}{10}$ of the figure. So, if one of the hexagons can be made using one yellow block, one red block, and one green block, an addition sentence for this could be written as $\frac{6}{10} + \frac{3}{10} + \frac{1}{10} = 1$ if the entire region is thought of in terms of green blocks. This means that $\frac{6}{10}$ of the region is yellow, $\frac{3}{10}$ is red, and $\frac{1}{10}$ is green.

Concentric Hexagons *(pages 54-56)*

When you see a dart board with a very small circle in the center and then several proportionally larger rings around it, those are concentric circles. The hexagons formed on pages 54, 55, and 56 with a yellow block in the center will resemble a six-sided dart board. Encourage students to make concentric geometric figures on grid paper, color them and discuss them. Display the figures on the bulletin board.

Puzzling Parallelograms *(page 57)*

It will take six red blocks for the parallelogram at the top of the page using the fewest number of blocks. The fraction sentence for this is $\frac{6}{6} = 1$. For the second parallelogram, it will take eighteen green blocks and the fraction sentence can be expressed as $\frac{18}{18} = 1$. The last parallelogram has numerous solutions. Have students work in pairs or small groups to find as many solutions as possible and record them in a chart. Extend this activity by asking students to find 2-color and 4-color solutions.

Puzzling Parallelograms (continued)

Number of Blocks	Number of Each Color				Fraction Sentence				
Number	Yellow	Red	Blue	Green	Yellow	Red	Blue	Green	
6	2		2	2	$\frac{12}{18}+$		$\frac{4}{18}+$	$\frac{2}{18}$	=1
8	2		1	4	$\frac{12}{18}+$		$\frac{2}{18}+$	$\frac{4}{18}$	=1
7	1	3		3	$\frac{6}{18}+$	$\frac{9}{18}+$		$\frac{3}{18}$	=1
9	1	2		6	$\frac{6}{18}+$	$\frac{6}{18}+$		$\frac{6}{18}$	=1
11	1	1		9	$\frac{6}{18}+$	$\frac{3}{18}+$		$\frac{9}{18}$	=1
7		5	1	1		$\frac{15}{18}+$	$\frac{2}{18}+$	$\frac{1}{18}$	=1
8		4	2	2		$\frac{12}{18}+$	$\frac{4}{18}+$	$\frac{2}{18}$	=1
9		4	1	4		$\frac{12}{18}+$	$\frac{2}{18}+$	$\frac{4}{18}$	=1
8		3	4	1		$\frac{9}{18}+$	$\frac{8}{18}+$	$\frac{1}{18}$	=1
9		3	3	3		$\frac{9}{18}+$	$\frac{6}{18}+$	$\frac{3}{18}$	=1
10		3	2	5		$\frac{9}{18}+$	$\frac{4}{18}+$	$\frac{5}{18}$	=1
11		3	1	7		$\frac{9}{18}+$	$\frac{2}{18}+$	$\frac{7}{18}$	=1
9		2	5	2		$\frac{6}{18}+$	$\frac{10}{18}+$	$\frac{2}{18}$	=1
10		2	4	4		$\frac{6}{18}+$	$\frac{8}{18}+$	$\frac{4}{18}$	=1
11		2	3	6		$\frac{6}{18}+$	$\frac{6}{18}+$	$\frac{6}{18}$	=1
12		2	2	8		$\frac{6}{18}+$	$\frac{4}{18}+$	$\frac{8}{18}$	=1
13		2	1	10		$\frac{6}{18}+$	$\frac{2}{18}+$	$\frac{10}{18}$	=1
9		1	7	1		$\frac{3}{18}+$	$\frac{14}{18}+$	$\frac{1}{18}$	=1
10		1	6	3		$\frac{3}{18}+$	$\frac{12}{18}+$	$\frac{3}{18}$	=1
11		1	5	5		$\frac{3}{18}+$	$\frac{10}{18}+$	$\frac{5}{18}$	=1
12		1	4	7		$\frac{3}{18}+$	$\frac{8}{18}+$	$\frac{7}{18}$	=1
13		1	3	9		$\frac{3}{18}+$	$\frac{6}{18}+$	$\frac{9}{18}$	=1
14		1	2	11		$\frac{3}{18}+$	$\frac{4}{18}+$	$\frac{11}{18}$	=1
15		1	1	13		$\frac{3}{18}+$	$\frac{2}{18}+$	$\frac{13}{18}$	=1

Two-Color Hexagons (*pages 58-59*)

These two pages are similar to the concentric hexagon pages. The fewest number of blocks is used to make each hexagon with only two colors. Remind students to approach these problems the same way as those on pages 54-56 using the same line of questioning. On page 58, it takes seven yellow and six blue blocks to make the hexagon showing that $\frac{21}{27}$ (or $\frac{7}{9}$) is yellow and $\frac{6}{27}$ (or $\frac{2}{9}$) is blue. On page 59, it takes thirteen yellow and six red blocks to make the hexagon showing $\frac{26}{32}$ (or $\frac{13}{16}$) is yellow and $\frac{6}{32}$ (or $\frac{3}{16}$) is red. At the bottom of each page, students are challenged to find the greatest number of blocks in two or three colors.

Name Parts of a Set

Cover each figure with a pattern block. You may need to move or remove various pattern blocks as you find the answers to the questions below.

1. How many blocks are there in all?_____

   What fraction of the set is each block?_____

2. How many of the blocks are blue?_____

   What fraction of the set is blue?_____

3. How many of the blocks are orange?_____

   What fraction of the set is orange?_____

4. How many of the blocks are blue or green?_____

   What fraction of the set is blue or green?_____

5. How many of the blocks are triangles?_____

   What fraction of the set are triangles?_____

6. How many of the blocks are trapezoids?_____

   What fraction of the set are trapezoids?_____

7. How many of the blocks are squares?_____

   What fraction of the set are squares?_____

8. How many of the blocks are regular polygons?_____

   What fraction of the set are regular polygons?_____

9. How many of the blocks are quadrilaterals?_____

   What fraction of the set are quadrilaterals?_____

10. How many of the blocks have right angles?_____

    What fraction of the set have right angles?_____

11. How many of the blocks have more than four sides?_____

    What fraction of the set has more than four sides?____

Think of more questions to ask about this set. Challenge a classmate to solve them.

Geometry and Fractions with Pattern Blocks
©Learning Resources, Inc.

Name Parts of Another Set

Cover each figure below with a pattern block. You may need to move or remove various pattern blocks as you find the answers to the questions below.

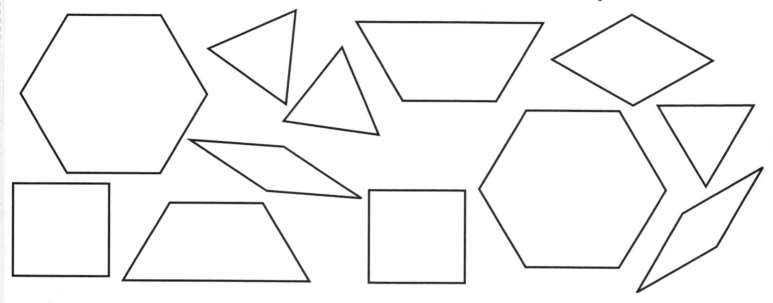

1. How many blocks are there in all?_____

   What fraction of the set is each block?_____

2. How many of the blocks are hexagons?_____

   What fraction of the set are hexagons?_____

3. How many of the blocks are triangles?_____

   What fraction of the set are triangles?_____

4. How many of the blocks are rhombuses?_____

   What fraction of the set are rhombuses?_____

5. How many of the blocks are tan?_____

   What fraction of the set is tan?_____

6. How many of the blocks are green or red?_____

   What fraction of the set is green or red?_____

7. How many of the blocks have four sides?_____

   What fraction of the set has four sides?_____

8. How many of the blocks have fewer than six sides?_____

   What fraction of the set has fewer than six sides?_____

9. How many of the blocks have right angles?_____

   What fraction of the set has right angles?_____

10. How many of the blocks are quadrilaterals?_____

    What fraction of the set are quadrilaterals?_____

11. How many of the blocks are regular polygons?_____

    What fraction of the set are regular polygons?_____

Think of more questions to ask about this set. Challenge a classmate to solve them.

Find Missing Parts of a Set

Cover each figure below with a pattern block. Add more pattern blocks to find the answer to each question. There may be more than one solution to each question.

 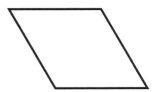 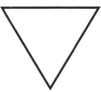 

1. To show $\frac{5}{8}$ of the blocks are triangles, add _____

2. To show $\frac{2}{5}$ of the blocks are trapezoids, add _____

3. To show $\frac{1}{2}$ of the blocks are squares, add _____

4. To show $\frac{1}{3}$ of the blocks are triangles, add _____

5. To show $\frac{1}{2}$ of the blocks are green, add _____

6. To show $\frac{3}{5}$ of the blocks are quadrilaterals, add _____

7. To show $\frac{3}{5}$ of the blocks are rhombuses, add _____

8. To show $\frac{1}{3}$ of the blocks are hexagons, add _____

9. To show $\frac{1}{2}$ of the blocks are tan, add _____

10. To show $\frac{1}{4}$ are trapezoids and $\frac{1}{2}$ are triangles, add _____

11. To show $\frac{1}{3}$ are blue and $\frac{1}{3}$ are orange, add _____

12. To show $\frac{1}{3}$ are squares and $\frac{1}{2}$ are triangles, add _____

Think of more questions to ask about this set. Challenge a classmate to solve them.

Geometry and Fractions with Pattern Blocks
©Learning Resources, Inc.

Find Missing Parts of Another Set

Cover each figure below with a pattern block. Add more pattern blocks to find the answer to each question. There may be more than one solution to each question.

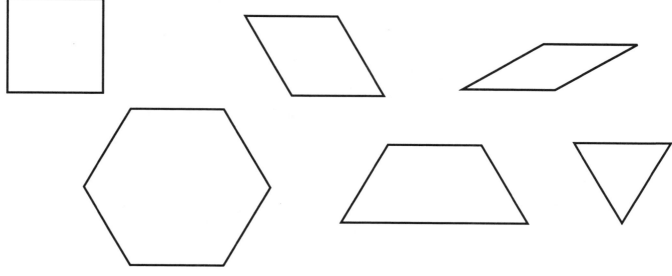

1. To show $\frac{5}{10}$ of the blocks are green, add _____

2. To show $\frac{1}{2}$ of the blocks are hexagons, add_____

3. To show $\frac{3}{8}$ of the blocks are squares, add _____

4. To show $\frac{1}{8}$ of the blocks are tan, add_____

5. To show $\frac{2}{7}$ of the blocks are trapezoids, add _____

6. To show $\frac{7}{12}$ of the blocks are triangles, add _____

7. To show $\frac{1}{2}$ of the blocks are quadrilaterals, add _____

8. To show $\frac{1}{5}$ of the blocks are triangles, add_____

9. To show $\frac{1}{3}$ of the blocks are blue, add _____

10. To show $\frac{5}{9}$ of the blocks are **not** quadrilaterals, add _____

Think of more questions to ask about this set. Challenge a classmate to solve them.

Fractional Parts of Each Block

1. Make this shape with green blocks.
   What fractional part of the shape is each green block?_____

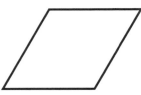

2. Make this shape with green blocks.
   What fractional part of the shape is each green block?_____

3. Make this shape with green blocks.
   What fractional part of the shape is each green block?_____

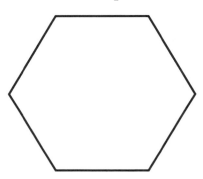

4. What did you notice about the value of the green block in each of the problems above?_____

5. Make this shape with red blocks.
   What fractional part of the shape is each red block?_____
   Make the shape with blue blocks.
   What fractional part of the shape is each blue block?_____

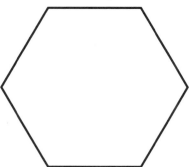

Geometry and Fractions with Pattern Blocks
©Learning Resources, Inc.

Parts of a Hexagon

Cover the shaded part of each hexagon with green blocks. Then write the fraction for the number of green blocks used.

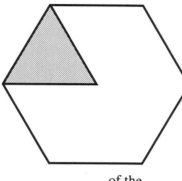

_____ of the
shape is green

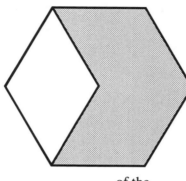

_____ of the
shape is green

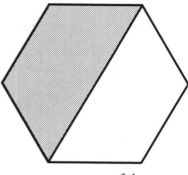

_____ of the
shape is green

Cover the shaded part of each hexagon with blue blocks. Then write the fraction for the number of blue blocks used.

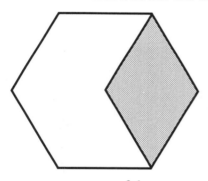

_____ of the
shape is blue

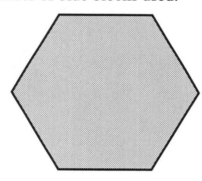

_____ of the
shape is blue

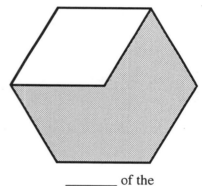

_____ of the
shape is blue

Cover the shaded part of the hexagon with red blocks. Then write the fraction for the number of red blocks used.

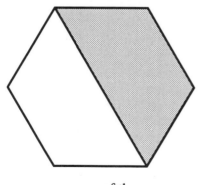

_____ of the
shape is red.

Sums of One on a Hexagon

Cover each hexagon differently. Use two, three, four, five, or six blocks. Trace and color the blocks used for each hexagon. Write an addition sentence.

Example:

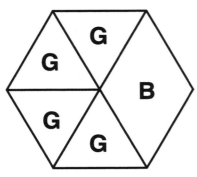

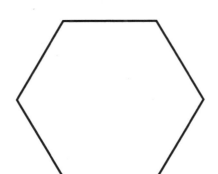

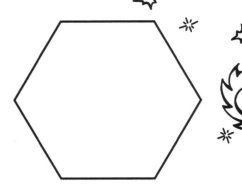

If you use one blue and four green blocks, then $\frac{1}{3}$ is blue and $\frac{4}{6}$ is green.

_____=1

_____=1

_____=1

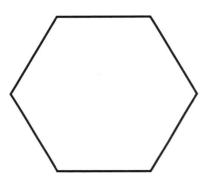

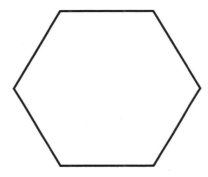

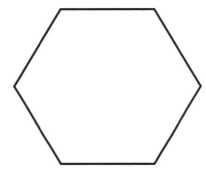

_____=1

_____=1

_____=1

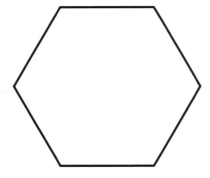

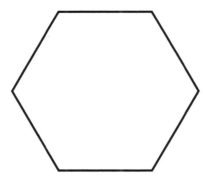

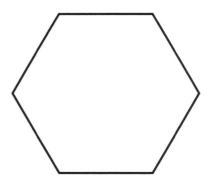

_____=1

_____=1

_____=1

Fractional Parts of a Region

Cover, trace and color each figure to show the fraction.

Show $\frac{1}{4}$

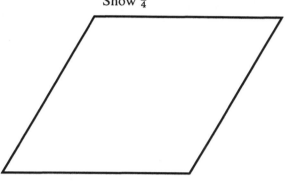

Show $\frac{3}{4}$

Show $\frac{2}{4}$

Show $\frac{5}{6}$

Show $\frac{3}{4}$

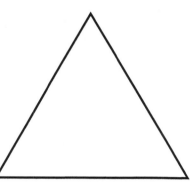

Show $\frac{6}{6}$

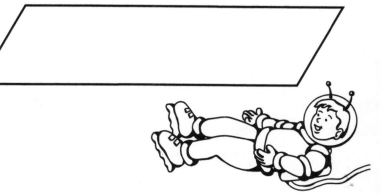

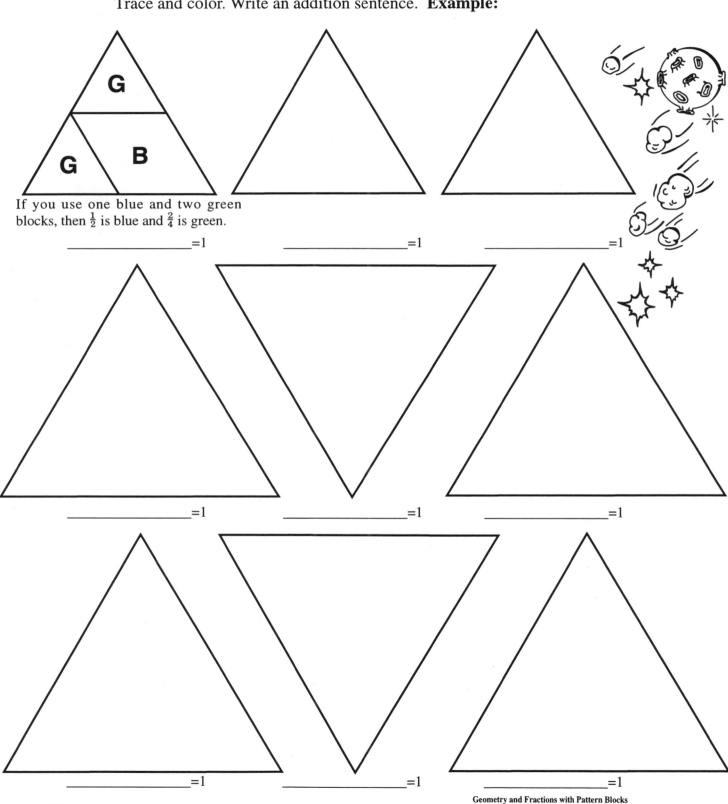

Sums of One on a Triangle

Cover each triangle differently using green, blue, red, or yellow blocks.
Trace and color. Write an addition sentence. **Example:**

G

G   B

If you use one blue and two green
blocks, then $\frac{1}{2}$ is blue and $\frac{2}{4}$ is green.

_____=1   _____=1   _____=1

_____=1   _____=1   _____=1

_____=1   _____=1   _____=1

Geometry and Fractions with Pattern Blocks
©Learning Resources, Inc.

Colorful Fraction Hexagons

Cover each hexagon differently using green, blue, red, or yellow blocks.
Trace and color. Write an addition sentence.

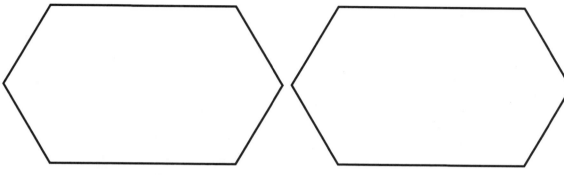

Use three blocks.

_____ = 1

Use three blocks.

_____ = 1

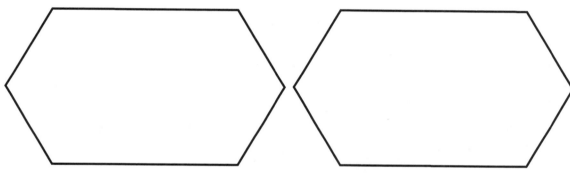

Use four blocks.

_____ = 1

Use four blocks.

_____ = 1.

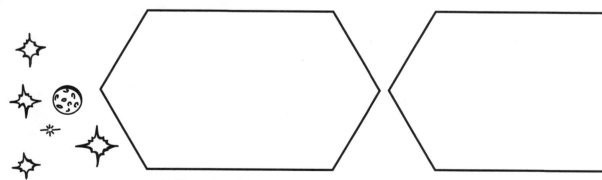

Use five blocks.

_____ = 1

Use six blocks.

_____ = 1

Concentric Hexagons

Make this hexagon with a yellow center and a red border. Trace and color the blocks to show your solution.

How many yellow blocks?_____
How many red blocks?_____
If you covered this shape with only red blocks, how many would you use?_____
What fractional part of the shape is yellow?_____
What fractional part of the shape is red?_____

Make this hexagon with a yellow center and a blue border. Trace and color the blocks to show your solution.

How many yellow blocks?_____

How many blue blocks?_____

If you covered this shape with only blue blocks, how many would you use?_____

What fractional part of the shape is yellow?_____

What fractional part of the shape is blue?_____

Another Concentric Hexagon

Make this hexagon with a yellow center, a blue inner border, and a red outer border. Trace and color the blocks to show your solution.

How many yellow blocks?_____

How many blue blocks?_____

How many red blocks?_____

If you covered this shape with only blue blocks,

how many would you use?_____

What fractional part of the shape is yellow?_____

What fractional part of the shape is blue?_____

What fractional part of the shape is red?_____

A Giant Concentric Hexagon

Make this hexagon with a yellow center, a green inner border, a blue middle border, and a red outer border. Trace and color.

How many yellow blocks?_____   How many green blocks?_____

How many blue blocks?_____   How many red blocks?_____

If you covered this shape with only green blocks, how many would you use?_____

What fractional part of the shape is yellow?_____

What fractional part of the shape is green?_____

What fractional part of the shape is blue?_____

What fractional part of the shape is red?_____

Geometry and Fractions with Pattern Blocks
©Learning Resources, Inc.

Puzzling Parallelograms

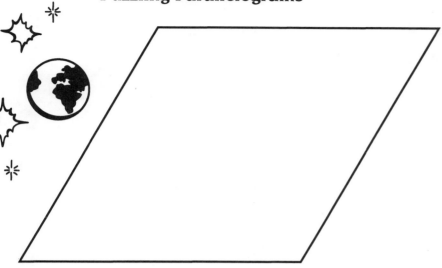

Make this parallelogram using the *fewest* number of same color blocks. Trace and color to show your solution. Write a fraction sentence for it.

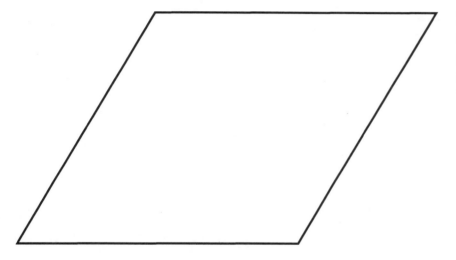

Make this parallelogram using the *greatest* number of same color blocks. Trace and color to show your solution. Write a fraction sentence for it.

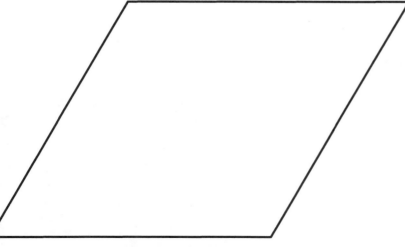

Make this parallelogram using three colors of blocks. Trace and color to show your solution. Write a fraction sentence for it.

Yellow and Blue Hexagon

Use thirteen yellow and blue pattern blocks to make this hexagon.

Trace and color the blocks to show your solution. Write an addition sentence for the shape.

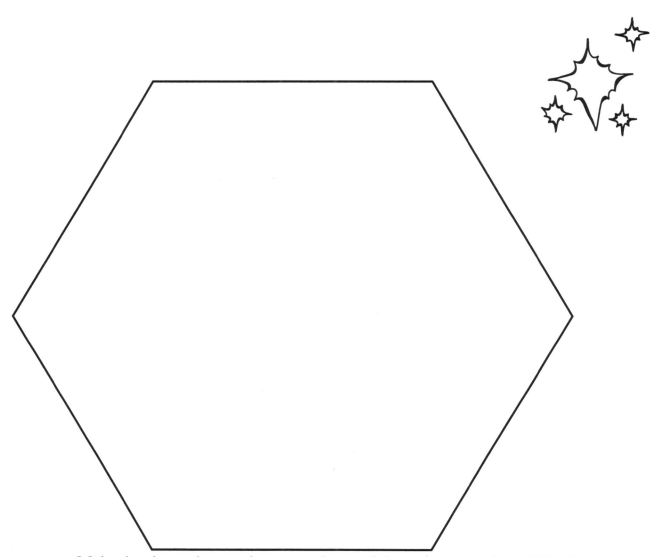

Make the shape above using two colors and the greatest number of blocks.

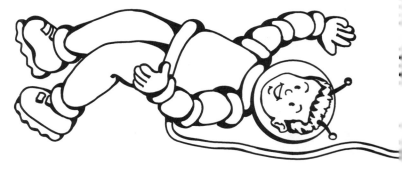

Yellow and Red Hexagon

Use nineteen yellow and red pattern blocks to make this hexagon. Trace and color the blocks to show your solution. Write an addition sentence for the shape.

Make the shape above using three colors and the fewest number of blocks.

Selected Solutions

Note: At least one solution is provided for each problem or exercise, many others are possible.

Geometry with Pattern Blocks

Page 15: Triangle: A closed figure with three sides; green triangle. Hexagon: A closed figure with six sides; yellow hexagon. Quadrilateral: A closed figure with four sides; orange square, blue rhombus, tan rhombus, red trapezoid. Regular Polygon: A closed figure that is equilateral (congruent sides) and equiangular (congruent angles); orange square, green triangle, yellow hexagon.

Page 16 Parallelogram: A quadrilateral formed by two pairs of parallel lines; orange square, blue rhombus, tan rhombus. Rhombus: A quadrilateral formed by two pairs of parallel lines is equilateral; orange square, blue rhombus, tan rhombus. Rectangle: A quadrilateral formed by two pairs of parallel lines and contains all right angles; square. Square: A quadrilateral formed by two pairs of parallel lines, equilateral, and contains all right angles; orange square. The orange square fits all the attributes listed for each term. The blue and tan rhombuses cannot be described as a special rectangle or square because they do not contain right angles.

Page 17:

Page 18:

Page 19:

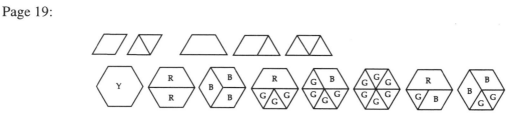

Page 20:

Number of Blocks	Number of each color			
	Y	R	B	G
4	1			3
3		3		
4		2	1	1
5		2		3
5		1	2	2
6		1	1	4
7		1		6
6			3	3
7			2	5
8			1	7
9				9

Page 21:

Number of Blocks	Number of each color			
	Y	R	B	G
3	1			2
3		2	1	
4		2		2
4		1	2	1
5		1	1	3
6		1		5
4			4	
5			3	2
6			2	4
7			1	6
8				8

Page 22:

Number of Blocks	Number of each color			
	Y	R	B	G
4	1	1	1	1
5	1	1		3
5	1		2	2
6	1		1	4
5		3	1	1
6		2	2	2
7		2	1	4
6		1	4	1
7		1	3	3
8		1	2	5
9		1	1	7

Page 23:

Number of Blocks	Number of each color			
	Y	R	B	G
7	3		2	2
8	3		1	4
7	2	2	3	
9	2		5	2
10	2		4	4
11	2		3	6
12	2		2	8
13	2		1	10
8	1	4	3	
9	1	2	6	
10		6	2	2
11		6	1	4
11		4	5	2
12		4	4	4
13		4	3	6
14		4	2	8
15		4	1	10
12		2	8	2
13		2	7	4
14		2	6	6
15		2	5	8
16		2	4	10
17		2	3	12
18		2	2	14
19		2	1	16

Page 24:

Page 25

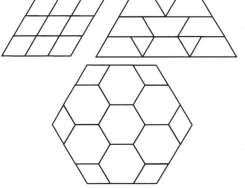

Page 26: Makes triangles with 3, 4, or 5 units on each side.

Page 27: Make parallelograms have 3 and 6 units in the sides, and 4 and 8 units on the sides; The longer side is twice the length of the short side on the parallelograms.

Page 28:

Shape	Area	Perimeter
1 by 1	1	4
2 by 2	4	8
3 by 3	9	12
4 by 4	16	16
5 by 5	25	20
6 by 6	36	24

Page 29:

Each Side	Area	Perimeter
1 unit	1	3
2 units	4	6
3 units	9	9
4 units	16	12
5 units	25	15
6 units	36	18

Page 30:

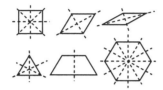

Page 31:

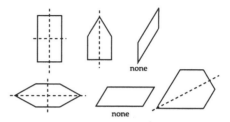

Page 32:

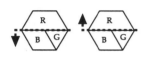

Page 33:

Pages 34-37: Refer to Teaching Notes on page 14.

Page 38: Row **1**) 90 degrees, 120 degrees, 60 degrees Row **2**) 60 degrees, 30 degrees, 120 degrees Row **3**) 120 degrees, 60 degrees, 120 degrees

Page 39:

| Sum=360° | Sum=180° | Sum=360° | Sum=360° | Sum=360° | Sum=720° |

1) 180 degrees, **2**) 360 degrees, **3**) 720 degrees, **4**) 540 degrees, **5**) 1080 degrees

Page 40:

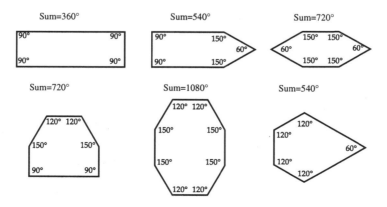

Fractions with Pattern Blocks

Page 44:

   1) 8 and $\frac{1}{8}$ **2)** 2 and $\frac{1}{4}$ **3)** 3 and $\frac{3}{8}$ **4)** 4 and $\frac{1}{2}$ **5)** 2 and $\frac{1}{4}$

   6) 1 and $\frac{1}{8}$ **7)** 3 and $\frac{7}{8}$ **8)** 5 and $\frac{5}{8}$ **9)** 6 and $\frac{3}{4}$ **10)** 3 and $\frac{3}{8}$

   11) 0 and $\frac{0}{8}$

Page 45:

   1) 12 and $\frac{1}{12}$ **2)** 2 and $\frac{1}{6}$ **3)** 3 and $\frac{1}{4}$ **4)** 3 and $\frac{1}{4}$ **5)** 2 and $\frac{1}{6}$

   6) 5 and $\frac{5}{12}$ **7)** 7 and $\frac{7}{12}$ **8)** 10 and $\frac{5}{6}$ **9)** 2 and $\frac{1}{6}$ **10)** 7 and $\frac{7}{12}$

   11) 7 and $\frac{7}{12}$

Page 46:

   1) 4 triangles   **2)** 1 trapezoid   **3)** 2 squares   **4)** 1 triangle and any other block

   5) 2 triangles   **6)** triangle or hexagon   **7)** square or rhombus

   8) 2 hexagons   **9)** 4 tan blocks   **10)** 1 trapezoid and 3 triangles

   11) 2 blue, 2 orange, and 1 other block   **12)** 3 squares and 5 triangles

Page 47:

   1) 4 green   **2)** 4 hexagons   **3)** 2 squares   **4)** and other 2 non-tan blocks

   5) 1 trapezoid   **6)** 6 triangles   **7)** 2 triangles or 2 hexagons or 1 triangle and 1 hexagon

   8) 1 triangle and 3 other 4 or 6 sided blocks

   9) 2 blue and 1 other colored block   **10)** 3 triangles or 3 hexagons

Page 48:

   1) 2 green blocks, $\frac{1}{2}$ **2)** 3 green blocks, $\frac{1}{3}$ **3)** 6 green blocks, $\frac{1}{6}$

   4) The value of the green block changed in relation to the shape it was in.

   5) 2 red blocks, $\frac{1}{2}$; 3 blue blocks, $\frac{1}{3}$

Page 49:

   Row 1: $\frac{1}{6}, \frac{4}{6}$ or $\frac{2}{3}, \frac{3}{6}$ or $\frac{1}{2}$ Row 2: $\frac{1}{3}, \frac{3}{3}, \frac{2}{3}$ Row 3: $\frac{1}{2}$

Page 50:

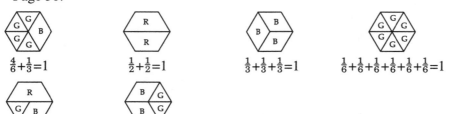

Page 51:

   Show $\frac{1}{4}$; cover with one blue block; Show $\frac{3}{4}$, cover with three orange blocks; Show $\frac{2}{4}$; cover with 2 tan blocks; Show $\frac{5}{6}$; cover with 5 orange blocks; Show $\frac{3}{4}$; cover with 3 green blocks; Show $\frac{6}{6}$; cover with 6 green blocks.

Page 52:

Page 52 continued:

$\frac{2}{3}+\frac{2}{9}+\frac{1}{9}=1$ $\frac{1}{3}+\frac{2}{9}+\frac{2}{9}+\frac{1}{9}+\frac{1}{9}=1$ $\frac{2}{9}+\frac{2}{9}+\frac{2}{9}+\frac{1}{9}+\frac{1}{9}+\frac{1}{9}+=1$ $\frac{1}{9}+\frac{1}{9}+\frac{1}{9}+\frac{1}{3}+\frac{1}{3}=1$ $\frac{1}{9}+\frac{1}{9}+\frac{1}{9}+\frac{1}{9}+\frac{1}{9}+\frac{1}{9}+\frac{1}{9}+\frac{1}{9}+\frac{1}{9}=1$

Page 53:

3 Blocks   3 Blocks   4 Blocks   4 Blocks   5 Blocks   6 Blocks

    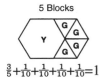

$\frac{3}{5}+\frac{1}{5}+\frac{1}{5}=1$ $\frac{3}{5}+\frac{3}{10}+\frac{1}{10}=1$ $\frac{3}{5}+\frac{1}{5}+\frac{1}{10}+\frac{1}{10}=1$ $\frac{3}{10}+\frac{3}{10}+\frac{1}{5}+\frac{1}{5}=1$ $\frac{3}{5}+\frac{1}{10}+\frac{1}{10}+\frac{1}{10}+\frac{1}{10}=1$ $\frac{3}{10}+\frac{1}{5}+\frac{1}{5}+\frac{1}{10}+\frac{1}{10}+\frac{1}{10}=1$

Page 54:   Page 55:

         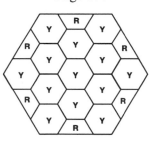

1 yellow $\frac{2}{8}$ or $\frac{1}{4}$ yellow   1 yellow $\frac{3}{12}$ or $\frac{1}{4}$ yellow
6 red $\frac{6}{8}$ or $\frac{3}{4}$ red   9 blue $\frac{9}{12}$ or $\frac{3}{4}$ blue
8 red   12 blue

1 yellow
9 blue
10 red
27 blue
$\frac{3}{27}$ or $\frac{1}{9}$ yellow
$\frac{9}{27}$ or $\frac{1}{3}$ blue
$\frac{15}{27}$ or $\frac{5}{9}$ red

Page 56:   Page 57:   Page 59:

See Teaching Notes
on pages 42 and 43.

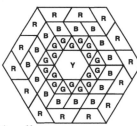   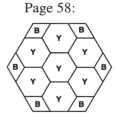

1 yellow
18 green
15 blue
14 red
96 green
$\frac{6}{96}$ or $\frac{1}{16}$ yellow
$\frac{18}{96}$ or $\frac{3}{16}$ green
$\frac{30}{96}$ or $\frac{5}{16}$ blue
$\frac{42}{96}$ or $\frac{7}{16}$ red

Page 58:

7 yellow
6 blue
27 blocks if all blue
$\frac{21}{27}+\frac{6}{27}=1$
or $\frac{7}{9}+\frac{2}{9}=1$

13 yellow
6 red
32 blocks if all red
$\frac{26}{32}+\frac{6}{32}=1$
$\frac{3}{16}+\frac{13}{16}=1$

Geometry and Fractions with Pattern Blocks
©Learning Resources, Inc.